Energy 73

旗帜飘飘

Fluttering Flags

Gunter Pauli

冈特·鲍利 著

凯瑟琳娜·巴赫 绘

高 芳 李原原 译

学林出版社
www.xuelinpress.com

丛书编委会

主　任：贾　峰

副主任：何家振　郑立明

委　员：牛玲娟　李原原　李曙东　吴建民　彭　勇
　　　　冯　缨　靳增江

丛书出版委员会

主　任：段学俭

副主任：匡志强　张　蓉

成　员：叶　刚　李晓梅　魏　来　徐雅清　田振军
　　　　蔡雩奇　程　洋

特别感谢以下热心人士对译稿润色工作的支持：

姜竹青　韩　笑　贾　芳　刘　晓　张黎立　刘之杰
高　青　周依奇　彭　江　于函玉　于　哲　单　威
姚爱静　刘　洋　高　艳　孙笑非　郑莉霞　周　蕊

目录

旗帜飘飘	4
你知道吗？	22
想一想	26
自己动手！	27
学科知识	28
情感智慧	29
艺术	29
思维拓展	30
动手能力	30
故事灵感来自	31

Contents

Fluttering Flags	4
Did you know?	22
Think about it	26
Do it yourself!	27
Academic Knowledge	28
Emotional Intelligence	29
The Arts	29
Systems: Making the Connections	30
Capacity to Implement	30
This fable is inspired by	31

蝙蝠倒挂在他最喜欢的树上,享受着喜马拉雅山脉的清风。他望着地平线,注意到在附近的山脊上,人们正在建造风力涡轮机来发电。

"我从来都弄不明白为什么风能在这里会如此受欢迎。"他对猫头鹰说。"我们该不会真的要牺牲山谷的美丽,就只为了发点电吧。"

A bat is hanging from his favourite tree, enjoying the fresh breeze of the Himalayas. He looks over the horizon and notices that on a mountain ridge nearby people are building wind turbines to generate electricity.

"I've never understood why wind energy is so popular here," he says to the owl. "We should really never have to sacrifice the beauty of our valley just to make some electricity."

享受着喜马拉雅山脉的清风

Enjoying the fresh breeze of the Himalayas

风力涡轮机的噪声

Noise of the wind turbines

"噢，风力涡轮机的噪声在我脑袋周围咆哮，真的让我很难受。"猫头鹰补充道，"幸运的是，这些新的风车转得比过去慢了。至少我们不会难受得想把头砍下来！"

"顺便问一下，你知道水的密度比空气大得多吗？"蝙蝠问。

"Oh, and the noise of the wind turbines snarling around my head really bothers me," adds the owl. "Fortunately, these new windmills turn slower than in the past. So at least we will not get our heads chopped off!"

"By the way, did you know that water is much denser than air?" asks the bat.

"那又怎样？"猫头鹰回答。

"嗯，速度是每小时15千米的水流，要比速度超过每小时300千米的飓风拥有更多的能量呢。"

"真的吗？那人们为什么不试着用冰川融化后丰富的水流产生能量呢？"猫头鹰问道。

"So?" replies the owl.
"Well, water flowing at 15 kilometres per hour has more power than a hurricane blowing at more than 300 kilometres per hour."
"Really? So why are people not trying to make energy out of the abundance of flowing water we get from the melting glaciers?" asks the owl.

为什么不用水流产生能量呢?

Why not make energy out of flowing water?

学会了如何建造巨大的水坝

Taught how to build huge dams

"问题是,人们仅仅学会了如何建造巨大的水坝来发电。不幸的是,这些水坝淹没森林和村庄,毁灭所有生物的生命,其实这么做是完全没有必要的。如果人们早知道会是这样……"

"嗯,木已成舟。不如再告诉我些关于水坝的事吧!人们不再需要它们了吗?"猫头鹰问道。

"The problem is that people were only taught how to build huge dams to generate electricity. Unfortunately, these dams can flood forests and villages, ruining the lives of all living things, without really having to do so. If only they had known…"

"Well, what's done is done. Rather tell me more about these dams. Are they not needed anymore?" asks the owl.

"在过去，这是人们能想出的最好的解决方法了。但是今天，你可以把涡轮机放到水管中发电。"蝙蝠说。

"那是怎么做到的呢？"猫头鹰问道。

"水流通过管道时，涡轮就会旋转，并把由此产生的能量传递到发电机中，从而产生电力。涡轮机垂直地放置在管道中，沿着管道旋转。"

"In the old days it was the best solution they could come up with. But today you can put turbines in a water pipe to generate electricity," says the bat.

"So how does that work?" asks the owl.

"When water flows through the pipes, the turbines spin and send the energy created by doing that to a generator, which makes electricity. The turbines are placed vertically in the pipe and turn along with the pipe."

把涡轮机放到水管中

Put turbines in a water pipe

把涡轮机放到管道中为电灯提供电力

Turbines into pipes to power the lights

"这么说,把它们垂直而不是水平放置,就能安装更多的涡轮机?"猫头鹰问道。

"是啊!"蝙蝠回答。

"好简单啊!为什么人们不用这个方法取代建更多的水坝呢?"猫头鹰惊呼道。

"哦,好多年前人们就这么做了。中国的家庭还把小涡轮机放到小管道中为电灯提供电力。现在,甚至整个城市都开始利用水流发电。这种方法正在流行起来。"

"So instead of placing them horizontally, they fit in more turbines by placing them vertically?" asks the owl.

"Yes!" answers the bat.

"So simple! Why don't people do this instead of building more dams?" exclaims the owl.

"Oh, it has been done for years. In Chinese homes they even put tiny turbines into small pipes to power the lights. Even whole cities are starting to produce power by using the flow of water. They are catching on."

"真是聪明。我听说制造这些大型风力涡轮机需要大量的磁铁和稀土金属,更不用说还需要那些巨大的塔架了。"

"嗯,这些涡轮机沿用了从中世纪传下来的利用旋转产生能量的思想。但我认为这个思想时间上可能比旋转木马还要早。旋转产生过多的摩擦,而只要有摩擦存在,一切最终都将归于结束。"

"That's smart. I hear these big wind turbines need lots of magnets and rare earth metals to work, not to mention those huge masts."

"Well, these turbines follow the old logic of the Middle Ages by making use of rotation. But I think the time has come to go beyond the concept of a merry-go-round. They cause too much friction and whenever there is friction, things eventually break."

需要大量的磁铁和稀土金属

Need lots of magnets and rare earth metals

让东西随风摆动

Let things flutter in the wind

"那么，有没有更好的解决方案呢?"猫头鹰很好奇。

"当然有！那就是让东西随风摆动。"

"就像用来祈祷的旗帜和船帆那样在风中摆动?"猫头鹰问道。

"So, is there a better solution then?" wonders the owl.

"Absolutely! You just let things flutter in the wind."

"Flutter in the wind like prayer flags and boat sails do?" asks the owl.

"完全正确，"蝙蝠热情地说，"你可以把一块磁铁放在桅杆上，再将另一块捆在固定旗子或船帆的绳子上。当旗子或船帆在风中飘扬时，将会产生电流。"

"太好了!"猫头鹰惊呼道。

"这个山谷祈祷的人越多，产生的电也就越多。"

……这仅仅是开始！……

"Exactly," says the bat enthusiastically. "You can just put a magnet on the mast and another one on the rope that holds the flag or sail in place and as it flutters in the wind, it will generate electricity."

"Wonderful!" exclaims the owl.

"The more prayers offered in this valley, the more electricity will be generated for its people."

… AND IT HAS ONLY JUST BEGUN!…

……这仅仅是开始！……

...AND IT HAS ONLY JUST BEGUN!...

Did You Know?
你知道吗？

因为风力涡轮机的叶片很长，比波音747的机翼还长，所以它们可能产生很大的噪声。

Because the blades of a wind turbine are very long – longer than the wingspan of a Boeing 747 – they have the potential to generate a lot of noise.

每年，全美国会有十亿只鸟撞到建筑物的窗户上，而汽车和卡车又会杀死另外十亿只鸟。小叶片的风车表面积小，却比具有大表面积的大叶片风车更容易杀死鸟类。

Every year, throughout the United States of America, a billion birds collide against the windows of buildings. Cars and trucks kill about another billion birds. Windmills with small blades, and therefore smaller surface areas, kill more birds than windmills with very big blades and large surface areas.

现代水力涡轮机可以把高达90%的可用能源转换成电能。在美国现有的80 000个水坝中只有2400个用于发电。

Modern hydro turbines can convert as much as 90% of available energy into electricity. Only 2 400 of the 80 000 existing dams in the United States are used to generate power.

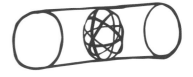

不丹王国只靠水力发电,并且75%的电力用于出口,占全国总出口的40%和GDP的25%。

The Kingdom of Bhutan produces electricity from hydropower alone and 75% of this electricity is exported. This makes up 40% of all exports and 25% of the country's GDP.

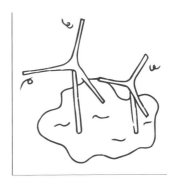

水磨自希腊和罗马时代就开始使用，转动磨盘使锤子发力来粉碎矿石。第一架风车于9世纪建于伊朗，12世纪才传入欧洲。

The watermill has already been used since Greek and Roman times to turn millstones and to power hammers to crush ore. The first windmill was constructed in the 9th century in Iran. Windmills were introduced to Europe in the 12th century.

藏传佛教的经幡（用来祈祷的旗帜）上四种表示尊贵的象征分别为：龙、金翅鸟（神话中的类鸟生物）、老虎和雪狮子。旗帜有五种颜色：蓝色、白色、红色、绿色和黄色，分别象征着天空、祥云、火焰、江河和大地。有人死了，就会升起108面经幡。

Tibetan Buddhist prayer flags display symbols of the four dignities: the dragon, the garuda (a mythical bird-like creature), the tiger, and the snow lion. The flags come in five colours: blue, white, red, green, and yellow, symbolising the sky, cloud, fire, water, and earth, respectively. When someone dies, 108 prayer flags are raised.

抽水马桶和厨房水龙头里的水流都可以通过一个安装在直径2毫米管道中的小涡轮机来发电。

The flow of water, from flushing toilets to kitchen taps, can be guided through tiny turbines fitted into a pipe as small as 2 mm in diameter to generate electricity.

类似于一个气球与衣服摩擦后就可以粘在墙上，一面旗帜的摆动也能产生能量，因为电荷具有在两种材料之间转移的能力。

Similar to how a balloon sticks to the wall after being rubbed against clothing, the fluttering motion of a flag generates energy because of the ability of charge to transfer between two materials.

Think About It

想一想

Would you like your valley to be full of windmills, or would you prefer to have water flowing through underground pipes to generate power?

你是喜欢让你的山谷布满风车,还是更喜欢让水流通过地下管道来发电呢?

如果旗帜可以发电,你认为它能够满足所有人日常生活中所需要的电力吗?

If flags can generate power, do you think that would be enough to provide all the electricity people need in their daily lives?

If water is so much denser than wind, why do we invest more in wind power than in hydropower?

如果水比风的密度大很多,那为什么我们更多地投资风力而不是水力呢?

如果水流通过小管道可以发电,有没有可能在你的水龙头上安装一个传感器呢?它可以告诉你洗手的水是不是太热。

If the flow of water through small pipes can generate power, would it be possible to install a sensor in your tap, which can tell if the water is too hot to wash your hands?

Do It Yourself!
自己动手!

Research the 10 main renewable energy sources of the future by consulting engineers about it. Then ask children under the age of 10 what they think the best ways are to generate electricity. Before you ask the children, tell them about the whale powering its own heart, leaves fluttering in the wind to aid trees in their absorption of CO_2, and the trout that swims against the current. Make a list of all the possible solutions put forward by the children and debate the differences. Who will be shaping the future? What will the energy sources of the future be?

通过咨询相关工程师，研究10种未来主要的可再生能源。问问10岁以下的孩子，他们所认为的最好的发电方式是什么。在问孩子们之前，告诉他们鲸用自己的心脏来提供动力，叶子在风中摆动来帮助树木吸收二氧化碳，以及鲑鱼逆流游泳。列出孩子们提出的所有可能的解决方案并且讨论其差异。谁将会塑造未来? 未来的能源将会是什么?

TEACHER AND PARENT GUIDE

学科知识
Academic Knowledge

生物学	树叶摆动可促进树木吸收二氧化碳，一棵树上在树冠顶部飘动的树叶比下面静止的树叶温度低2~4℃；树叶飘动的结果是使二氧化碳的传递加倍。
化 学	钕和钐这两种稀土金属可提供永久性的磁场。
物 理	很多物理元件的质量、体积和密度都是可以测量的；水不能被压缩，但空气可以；密度随压力或温度变化而变化；理想气体定律可以用来计算干空气、水蒸气或两者混合物的密度；阿基米德原理：浸入液体中的物体受到向上的浮力，浮力的大小等于物体所排开的液体受到的重力；声音在水中的传播速度比在空气中快，这是由于水的密度更高，允许更多的粒子碰撞；热在水中的损耗速度比在空气中快20倍，这是由于水的密度较高，热传递更快。
工程学	永久磁场的产生；1克每立方厘米(g/cm^3)和1千克每立方米(kg/m^3)是常见的密度单位，$1g/cm^3$等于$1000kg/m^3$；工程师们用钕磁铁材料取代了风力涡轮机组中的齿轮箱；滚珠轴承降低了摩擦；风力涡轮机的输出已经从过去的200千瓦/转提高成现在的7.5兆瓦/转。
经济学	水电比核能便宜50%，是化石燃料成本的40%，天然气成本的25%。
伦理学	人们宁愿沿着山脊线建水坝或风车而牺牲山谷的美丽，也不愿将水通过带有涡轮机的管道来保护山谷；不丹王国保护了52%的土地和生物多样性，但却没有保护河流，它们都被水坝利用了。
历 史	阿基米德通过排开水的多少计算黄金的体积；风车于9世纪在伊朗发明，12世纪由十字军带到欧洲，同时期的中国发明了风驱动的提水设备。
地 理	喜马拉雅山脉来自梵文"喜马"(雪)和"拉雅"(住宅)，包括了世界十大山峰中的9个；喜马拉雅山脉源于印度板块和亚欧板块的碰撞，跨越五个国家(印度、尼泊尔、不丹、中国和巴基斯坦)；喜马拉雅山脉是冰雪的第三大沉积地，仅次于南极和北极；挪威99%的能源来自水力发电。
数 学	密度等于质量除以体积，水的密度是空气的784倍。
生活方式	经幡是永久性的，而生命在演化、终结，并被新的生命所取代；孩子们喜欢旋转木马，它是一种在圆形的旋转平台上放置很多座位的游乐设施。
社会学	经幡是用来促进和平、同情、力量和智慧的。祈祷者不是向神祷告，而是让风带着美好的愿望传播祝福和同情心。
心理学	摆动是一种混乱或兴奋的状态。
系统论	能量无处不在。问题不是缺乏能量，而是我们不知道周围所有的能量来源。

教师与家长指南

情感智慧
Emotional Intelligence

蝙 蝠　　蝙蝠为风车可能毁了他的美丽山谷而感到紧张。尽管了解了这个问题可能的解决方案,他仍在哀叹人们的无知,不懂得水比风的力量更强。蝙蝠指出,政策制定者和金融家坚持利用传统的水电资源和容忍环境破坏是非常愚昧的。蝙蝠不仅表示担忧,还公开分享他对创新的认识并指出这些创新已经成功实施的案例。蝙蝠热情地解释了每个人如何能实现这些创意,并鼓励合理搭配和团体意识。

猫头鹰　　猫头鹰抱怨风车产生的噪声。他对创新很好奇,因为他渴望找到良好的解决方案。猫头鹰表示沮丧,因为有很明显的现成的机会人们却没有利用。他希望去探索其他方法,寻找更好的解决方案。然而他很担心蝙蝠的方法,那需要使用大量的稀土金属来制造很多磁铁。通过模仿旗帜摆动来发电的想法,吸引了他全部的注意力,因为这结合了功能(电)与精神(祈祷和平)的需求,所以他很感激他的朋友蝙蝠。

艺术
The Arts

很多人利用风吹过树木和水流经森林的声音来帮助他们放松。找到这些声音的录音并想象这些来源中可能产生的能量形式。让艺术激发科学吧!

TEACHER AND PARENT GUIDE

思维拓展
Systems: Making the Connections

我们周围存在很多的能量，它无时无刻无处不在。问题是，通常人们尤其是工程师们并没有注意到这一点，很少有人受过训练去寻找现存的小型本地能源。每个人都更关注大型、集中发电的方法，这种方法产生的电要通过覆盖数千公里的大型电网来传输。树木为在当地如何产生电力提供了一个很好的启发。通过模仿巧妙而简单的树叶摆动产生的电，可以与小型风力涡轮机的发电能力相媲美。风车的圆周运动、齿轮箱和涡轮都会产生摩擦和衰变，因此需要长期维护。可以用一种没有齿轮箱或涡轮的动力系统来取代它。这种动力系统产生的电足以满足当地的需求。大自然拥有用简单方法解决复杂问题的能力，利用本地分散的电力系统，比传统系统需要的基础设施、资本投资和后期维护更少。而方便易得的天然能源，只需要风力和水力就可运作。大到一个城市，小到一个家，每一幢建筑、每一所房子、每一间办公室都有水流入，但没人再次利用水的流出。如果工程师不能抓住这个机会利用可再生能源创造更多的能源，而是继续依靠不可再生能源的话，全球的电力需求将无法得到满足。相反，如果意识到了这个机会，我们便可以扭转局势，走向可持续发展。

动手能力
Capacity to Implement

我们不断地寻找新能源，但必须意识到能源就在我们周围，无时无刻无处不在。当不同的两种材料接触时，就会有电荷转移。如果让它们之间保持一定距离，就会产生电压，形成电流。因此，发电机可以将环境中的任意机械能转化为电能。尝试通过两种材料摩擦来发电，用一个LED小灯把它们连接起来。也许需要一点时间来产生足够的摩擦力，使产生的电量能点亮LED灯。当灯开始闪烁时，拍个照片和我们一起分享吧！

教师与家长指南

故事灵感来自

肖恩·弗拉伊内
Shawn Frayne

肖恩·弗拉伊内毕业于麻省理工学院，并获得物理学学位。2006年，他加入救援队帮助重建海地。他注意到当地人依赖煤油或柴油发电机，试图建一个便宜的风力发电机。他意识到风力涡轮机设计中正在取消齿轮箱，但涡轮技术仍然效率低下，尤其是使用范围很小时，因为所有技术都是为大规模使用而设计的。这推动了第一个无涡轮风力发电机——风箱的发明，其效率比燃气涡轮发动机高10~30倍。肖恩已经有了多项创新，并在香港的实验室组建了一支创意团队，通过黑线鳕（Haddock Invention）发明平台，在绿色包装、太阳能水消毒和能量储存方面进行了开创性工作。此外，他被《发现》杂志评为"40岁以下的20个最强大脑"之一，被《企业家》杂志评为"30岁以下30人"之一。

更多资讯

http://www.engineeringtoolbox.com/density-air-d-680.html

http://techxplore.com/news/2014-09-fluttering-flags-harvest-power.html

http://www.rexresearch.com/frayne/frayne.htm

www.lookingglassfactory.com

www.haddockinvention.com

图书在版编目（CIP）数据

旗帜飘飘：汉英对照／（比）冈特·鲍利著；
（哥伦）凯瑟琳娜·巴赫绘；高芳，李原原译．——上海：
学林出版社，2016.6
（冈特生态童书．第三辑）
ISBN 978-7-5486-1064-9

Ⅰ．①旗… Ⅱ．①冈… ②凯… ③高… ④李… Ⅲ．
①生态环境－环境保护－儿童读物－汉、英 Ⅳ．
① X171.1-49

中国版本图书馆 CIP 数据核字 (2016) 第 126070 号

© 2015 Gunter Pauli
著作权合同登记号 图字 09-2016-309 号

冈特生态童书
旗帜飘飘

作　　者——	冈特·鲍利
译　　者——	高　芳　李原原
策　　划——	匡志强
责任编辑——	程　洋
装帧设计——	魏　来
出　　版——	上海世纪出版股份有限公司 学林出版社
	地　址：上海钦州南路81号　电话／传真：021-64515005
	网　址：www.xuelinpress.com
发　　行——	上海世纪出版股份有限公司发行中心
	（上海福建中路193号 网址：www.ewen.co）
印　　刷——	上海丽佳制版印刷有限公司
开　　本——	710×1020　1/16
印　　张——	2
字　　数——	5万
版　　次——	2016年6月第1版
	2016年6月第1次印刷
书　　号——	ISBN 978-7-5486-1064-9/G·399
定　　价——	10.00元

（如发生印刷、装订质量问题，读者可向工厂调换）